BUILDING BLOCKS

Math Matters!

MW01705734

MEASURING in Our World

By Naomi Osborne

Cavendish Square
New York

When you measure something, you find out its size or how much of it there is.

Length is
how long something is.
You can measure length
with a tape measure.

2 3 3m 4 5 6 7 8 9 10

The United States measures length using inches and feet. There are 12 inches in 1 foot.

12 inches = 1 foot

Other countries measure length using centimeters and meters. There are 100 centimeters in 1 meter.

200 centimeters = 2 meters

Height measures
how tall something is.
A nurse measures
our height.

PTEOR
FZBDUE
BFLTCA
TPOELZD

As we grow,
our height changes.

Weight is how
heavy something is.
We can measure weight
in pounds or kilograms.

0 kg 10 20 30 40 50 60 kg 70 80 90 100 110
0 lb 20 40 60 80 100 120 140 lb 160 180 200 220 240

We use a scale to measure weight.

UNIT
Z/T
ON/OFF

People measure
food while cooking.

Measuring cups help us measure in the kitchen.

1/2 LITRE
450 ml
400 ml
350 ml
300 ml
1/4 LITRE
200 ml
150 ml
100 ml
50 ml
1 PINT
18 oz
16 oz
14 oz
12 oz
1/2 PINT
8 oz
6 oz
4 oz
2 oz

Measuring can be fun!

Published in 2021 by Cavendish Square Publishing, LLC
243 5th Avenue, Suite 136, New York, NY 10016

First Edition

Website: cavendishsq.com

Library of Congress Cataloging-in-Publication Data
Names: Osborne, Naomi (Children's nonfiction author), author.
Title: Measuring in our world / Naomi Osborne.
Description: First edition. | New York : Cavendish Square Publishing, 2021. | Series: Math matters!
Identifiers: LCCN 2020003328 (print) | LCCN 2020003329 (ebook) | ISBN 9781502656551 (library binding) | ISBN 9781502656537 (paperback) | ISBN 9781502656544 (set) | ISBN 9781502656568 (ebook)
Subjects: LCSH: Measurement–Juvenile literature.
Classification: LCC QA465 .O83 2021 (print) | LCC QA465 (ebook) | DDCo 530.8–dc23
LC record available at https://lccn.loc.gov/2020003328
LC ebook record available at https://lccn.loc.gov/2020003329

Editor: Vanessa Oswald
Copy Editor: Nathan Heidelberger
Designer: Deanna Paternostro

The photographs in this book are used by permission and through the courtesy of: Cover (top) Igor Pushkarev/Shutterstock.com; cover (bottom) John Arehart/Shutterstock.com; p. 3 LightField Studios/Shutterstock.com; p. 5 Seregam/Shutterstock.com; p. 7 Nigel Stripe/Shutterstock.com; p. 9 Josep M Penalver Rufas/Shutterstock.com; p. 11 CandyBox Images/Shutterstock.com; p. 13 Africa Studio/Shutterstock.com; p. 15 VGstockstudio/Shutterstock.com; p. 17 bestv/Shutterstock.com; p. 19 Monkey Business Images/Shutterstock.com; p. 21 M. Unal Ozmen/Shutterstock.com; p. 23 Syda Productions/Shutterstock.com.

CPSIA compliance information: Batch #CS20CSQ: For further information contact Cavendish Square Publishing LLC, New York, New York, at 1-877-980-4450.

Printed in the United States of America